The
Instant Physicist

The
Instant Physicist

Richard A. Muller

Illustrations by Joey Manfre

W. W. NORTON & COMPANY

NEW YORK ⚛ LONDON

For information about permission to reproduce selections from this book,
write to Permissions, W. W. Norton & Company, Inc.,
500 Fifth Avenue, New York, NY 10110

For information about special discounts for bulk purchases, please contact
W. W. Norton Special Sales at specialsales@wwnorton.com or 800-233-4830

Manufacturing by Toppan Printing
Book design by Ellen Cipriano
Production manager: Julia Druskin

Library of Congress Cataloging-in-Publication Data

Muller, R. (Richard)
The instant physicist : an illustrated guide / Richard A. Muller ;
illustrations by Joey Manfre.—1st ed.
p. cm.
ISBN 978-0-393-07826-8 (hardcover)
1. Physics—Popular works. I. Title.
QC24.5.M848 2011
530—dc22

2010025739

W. W. Norton & Company, Inc.
500 Fifth Avenue, New York, N.Y. 10110
www.wwnorton.com

W. W. Norton & Company Ltd.
Castle House, 75/76 Wells Street, London W1T 3QT

1 2 3 4 5 6 7 8 9 0

For Elizabeth, Melinda, and Layla

INTRODUCTION

Learn astonishing facts!

Win arguments with friends and relatives!

Make crazy-sounding bets, and win money!

This book promises all that and more . . . you can become an *instant physicist*.

Okay, mastering this book might not get you a job at a national laboratory, but it might help you pass yourself off as a physicist at your next party (assuming you would want to). And with clever bets based on the paradoxical surprises you'll learn here, you could pay for the cost of this book within the first day.

And yet, this book is more serious than it looks. Read it, browse it, graze on it, enjoy it, and you'll learn more than you might think. Unexpected nourishment lurks inside. Physics is full of surprises, paradoxes, and contradictions. That's partly why physics is hard to learn. Nonintuitive anecdotes can help unravel the mysteries, and physics with a chuckle is physics more easily remembered. Every page of this book has information worth knowing. We live in a high-tech world in which many, if not most, major issues have a physics connection. Think about these key areas: global warming and alternative energy; weather satellites and spy satellites; nuclear bombs and nuclear power plants; tsunamis and hurricanes. Whether you are a world leader or just a voting citizen, you'll improve your judgment by knowing some key facts and numbers. And what better way to learn them than through humor?

Most of the amazing facts in this book are included because they reflect something important about our high-tech world. Some are included simply because they're amazing. Start with the art, read the caption, be confused, puzzle about it, see if you can guess the resolution—and then read the explanation on the facing page. This is an absolutely painless way to learn.

The physics anecdotes in this book are distilled from my course at Berkeley, which is informally titled "Physics for Future Presidents," and from the popular book with that same title. For the past two years, this course was voted "Best Class" on the Berkeley campus—an honor that I can't imagine exceeding. But the anecdotes alone would be relatively dry without the wonderful art of my illustrator, Joey Manfre, and his delightful and gentle sense of humor. So sit back, enjoy, and learn some physics the easy way.

— Richard A. Muller, 2010

The
Instant Physicist

THE PHYSICS: The atmosphere is radioactive because it is constantly being bombarded by cosmic rays. All atmospheric carbon contains about one part in a trillion of C-14, a radioactive form of carbon that is also called "radiocarbon." Plants get all their carbon from the atmosphere in the process of photosynthesis, so all plants are radioactive. Animals that eat plants, or eat other animals that eat plants, also are radioactive. Every living creature on Earth is radioactive.

When we die, the radioactivity begins to go away. Half is gone in about 5,700 years, the half-life of radiocarbon. We can tell how long a plant or animal has been dead if we can extract some carbon and see how much radioactivity is left.

For all the radioactivity to decay away takes about 50 half-lives. That's about 300,000 years. For more about this radioactivity, see page 12.

If you aren't radioactive, then you're dead—and
you've been dead for a long time.

THE PHYSICS: The U.S. government has decided that alcohol for human consumption must be made from "natural" materials, such as grains, grapes, or fruit. That regulation rules out alcohol made from petroleum. Such alcohol is chemically identical to natural alcohol and just as safe—there's no difference in taste—so why this rule? The reasons have to do with history—keeping alcohol more expensive (a goal of the anti-alcohol lobby) and minimizing competition (a goal of the liquor lobby).

If there's no chemical difference, how can we tell natural alcohol apart from alcohol made from petroleum? The U.S. Bureau of Alcohol, Tobacco, Firearms and Explosives, charged with enforcing the natural-alcohol rule, has only one reliable test: check for radioactivity.

Natural alcohol gets its carbon from plants; the plants got the carbon from atmospheric carbon dioxide. As explained on page 10, atmospheric carbon dioxide is radioactive because of the continued bombardment of cosmic rays—particles coming from space—that collide with nitrogen molecules and turn it into C-14, radiocarbon. Only one atom in a trillion carbons in the atmosphere is radiocarbon, but that's enough to be detectable. (I invented what is currently the most sensitive way to detect C-14, called "accelerator mass spectrometry.")

Petroleum was also made from atmospheric carbon, but it was buried hundreds of millions of years ago, isolated from the radioactive atmosphere. Radiocarbon has a half-life of about 5,700 years, and after 100 million years, there is nary an atom of C-14 left. So lack of radioactivity in alcohol is a dead giveaway that it's made from petroleum and not "natural" materials.

True, a bootlegger could get some C-14 and add it to illegal liquor. But that's beyond the skill set of most bootleggers.

Liquor and wine in the United States must be
radioactive to be legal. When tested, drinking
alcohol is required to have at least
400 radioactive decays per minute for
each 750 milliliters.

THE PHYSICS: The Earth is warmed by visible light from the sun, and cooled by radiating away its heat as infrared (or "IR") radiation. By itself, the solar warming isn't enough to keep the surface temperature above freezing. If we didn't have an atmosphere, the surface of the oceans would freeze solid. But the atmosphere of the Earth acts as a blanket, reflecting some of the IR radiation back down to the surface, making it warmer.

This phenomenon is analogous to the behavior of a greenhouse, in which the glass roof allows sunlight to enter but prevents heat from leaving. (In a greenhouse, the heat escapes mostly through convection, not IR radiation, but the result is the same.)

When people worry about global warming, it is not because gases such as carbon dioxide will suddenly turn the atmosphere into a greenhouse. The atmosphere already is a greenhouse. Rather, carbon dioxide enhances the greenhouse effect by absorbing more of the Earth's IR radiation.

The "normal" greenhouse effect of the atmosphere warms the Earth about 30°F. The enhanced greenhouse effect, from extra carbon dioxide in the atmosphere, might warm the Earth an additional 3–5°F by the end of the century, if the calculations are right. That's why people worry about global warming.

If not for the notorious greenhouse effect,
the entire surface of the Earth would
currently be frozen solid.

THE PHYSICS: You can determine the energy density of a lithium-ion battery by weighing it and reading its label. My computer battery weighs one pound, cost me $120, and, according to the label, stores 60 watt-hours of energy. That means an energy density of 60 watt-hours per pound, and a cost of $2 per watt-hour. In contrast, a pound of gasoline has an energy of about 5,000 watt-hours, or about 83 times more than the battery. Of course, the battery can be recharged, but your gas tank can be refilled. The salient point is that for the same energy stored, batteries must weigh a lot more.

Batteries do have an important advantage: electric energy can be used about 5 times more efficiently than gasoline. So in reality, you don't need 83 times more pounds of battery than gasoline, but only 16 times more.

The Tesla Roadster, the most famous of the new electric cars, carries 1,100 pounds of batteries—more than half a ton—making up about 44% of the total weight of the car. At a retail price of $120 per pound of battery, that would be $132,000 just for the batteries alone. The cost per mile of travel in an electric car comes not from refueling, but from replacing the batteries, which are guaranteed for only 100,000 miles. Tesla says it will replace dead batteries (they have a calendar life of about 3 years) for less than $30,000. But that's still nearly $10,000 per year just for batteries!

Thinking of converting your Prius to a plug-in? The same problem—expensive batteries that demand frequent replacement—plagues plug-in hybrids. A recent analysis by Consumers Union showed that after 3 years of operation, you will have saved $2,000 in gasoline costs, but you'll have to pay $10,000 to replace the batteries. (I own a Prius and will not be converting it to a plug-in.)

No wonder most of us are still driving gasoline-powered cars.

Gasoline has 83 times more energy per pound than expensive lithium-ion computer batteries have. The batteries in the most famous electric car, the Tesla Roadster, would cost $132,000 if you bought them at the current retail cost for computer batteries. But Tesla gives a discount.

THE PHYSICS:

Although a tsunami travels fast, it is a very long and gradual wave. Crest to trough (high to low point) may be 30 feet vertically, but the horizontal distance is typically 10 miles.

Many people first recognize that a tsunami is coming when the water at the beach begins to recede. That ebb of the water signals the arrival of the trough, which often comes first. The crest is moving toward shore at 300 miles per hour, but it is still 10 miles away. Even at its airplane-like speed, it will take 2 minutes to reach you. So you have 2 minutes to climb up 30 feet. Don't run *away*; run *up*, on a hillside or some other structure that won't be knocked over. Do that, and you'll be safe.

As the tsunami crest approaches, it looks as if the tide is coming in rapidly. That's because the slope of the wave is so gradual. A 30-foot rise in 2 minutes means an average rate of one foot every 4 seconds. The tsunami does not approach as a "wall of water" (as is often misreported—humorously so in the illustration), although small waves on top of such a high tide can look pretty big when the tide has brought the ocean into your living room.

Many people think the term "tidal wave" is incorrect, but that's not true. That phrase is based on the way the wave behaves: it comes in like a fast tide. The word "tsunami," preferred by some because it is not English, means "harbor wave." Just as tidal waves aren't caused by the tides, tsunamis aren't caused by harbors—so arguments over which term is "correct" are silly. The wave is called a harbor wave by the Japanese because most of the damage from small tsunamis is done in harbors, where ships moored to docks are damaged by the waves.

Even though a tsunami, sometimes called a
"tidal wave," typically travels 300 miles
per hour, you can outrun it.

THE PHYSICS: The laser was invented by Charles Townes, a physicist then at Columbia University. The first laser didn't use visible light; instead, it used microwaves, a low-frequency form of light that is used in radar, microwave ovens, and now in cell phones. Townes called his initial device a MASER, an acronym for "microwave amplification by stimulated emission of radiation." (The physics principle that made it possible is called "stimulated emission." Such radiation had been predicted to exist by Albert Einstein.)

Townes believed that soon he or one of his students would be able to make the device work with visible light. The acronym for that version of the device was identical to the one for the microwave version, but with "L" for "light" substituted for the "M" of "microwaves." That's how we got to the term "laser."

To Townes, the maser and the laser were the same device working on the same principle and differing only in frequency. He toyed with the idea of giving them a common acronym. Both light and microwaves are forms of electromagnetic radiation, so he thought maybe he should use "ER" instead of "M" or "L," and let the one word—"eraser"—cover every frequency.

But, with a smile, he rejected the acronym ERASER. (At least, he smiled when he told me this story.) Townes eventually won the Nobel Prize for his invention, and he is now an astrophysicist at Berkeley.

"Eraser" was one of the names originally
considered for the laser, but "laser" won out.

THE PHYSICS: Heat radiation is also called "infrared" (or just IR) radiation. It is emitted by anything that is warm or hot. It is the invisible light that is emitted by heat lamps. Our eyes can't see it, but our skin can sense its warming effect. Your lips are the most sensitive part of your body for detecting IR radiation. If no one else is around for the lip experiment, try putting something hot (like tea) near your lips and see if you notice the heat. Then put the same hot thing near your finger; it is much less sensitive.

For more on infrared radiation, see page 14.

Heat radiation, used by snakes to detect prey
and by mosquitoes to detect you, can also be
sensed by humans. Try the following experiment:
close your eyes and move your lips close to those
of another person. See if you can tell when
you're getting close. Be careful. There could
be unintended consequences if you touch.

THE PHYSICS: Both pit vipers and Stinger missiles use infrared (IR) sensors. When hunting at night, a pit viper can't see its prey very well, but it can detect its victim's radiated heat (IR radiation) in the special regions on the sides of its head known as "pits." Something warm in front?—strike! (Rattlesnakes are one kind of pit viper.)

Stinger missiles operate on a similar principle. They use IR sensors to find anything hot in the sky—which usually turns out to be an airplane's exhaust. (There's really nothing else that hot up there.) Airplanes try to fool Stinger missiles by dropping hot flares, hoping the missiles will follow the flares instead.

Some home video cameras claim to work in total darkness, but they actually have lamps on them that illuminate the subject with IR radiation, visible to the camera but not to the eye. And our military personnel use IR sensors to see at night; this practice is what led to the U.S. Army's slogan "we own the night."

Pit vipers and Stinger missiles use the same physics to locate their prey.

THE PHYSICS: The original name for MRI was "nuclear magnetic resonance," or NMR. But when patients heard the word "nuclear," they got scared. Maybe they mistakenly thought they would be exposed to radioactivity. In fact, NMR doesn't use radioactivity; it just applies magnetism and radio waves to shake the inner part of a hydrogen atom—the nucleus—in a way that allows the body's density to be mapped.

Because many patients were so afraid of anything "nuclear," scientists and doctors decided they had to change the name of the method to "magnetic resonance imaging," or MRI. The procedure didn't change—only the name. The word "resonance" refers to the induced shaking of the nucleus.

Magnetic resonance imaging, or MRI,
produces amazing medical images of the
tissue structure inside the human body. But MRI
once had a different name—a name that scared
away so many patients that scientists and
doctors had to change it.

THE PHYSICS: Plutonium is pretty toxic. If inhaled, the lethal dose is a few thousandths of a gram. That's much worse than most common poisons, including arsenic. Early commentators exaggerated this danger a bit and mistakenly called plutonium the "most toxic substance known to man." Soon this misinformation became an urban legend. Not everyone made that mistake; as far back as 1956, the *Guinness Book of Records* correctly gave botulinum toxin the honor of "most toxic." This is the chemical created by the botulism bacteria, sometimes found in homemade mayonnaise, and it's the active ingredient in Botox. It is a million times more lethal than plutonium. But urban legends are difficult to stop.

Because it's so toxic, Botox should be administered only by a physician, who can make sure that it is injected in tiny amounts—much less than a lethal dose—enough to paralyze the wrinkle muscles but not do any permanent damage.

Plutonium, often called "the most toxic substance known to man," is actually 1,000 times *less* toxic than Botox, the chemical that people inject under their skin to reduce wrinkles.

THE PHYSICS: You'll never get cancer from 2,500 rem, because you'll die so quickly. Yet 2,500 rem is called the "cancer dose." The reason is that cancer is a game of chance. Small doses put you at risk. If you spread 2,500 rem over 2,500 people, giving only one rem to each person, nobody will die of radiation illness—but on average, there will be one extra cancer among those 2,500 people. There's no way to guess which person will get it. So 2,500 rem will trigger for a single cancer—it is "one cancer dose"—provided it is shared widely enough to avoid immediate death from radiation illness.

Radiation damage to the body is measured in
"rem." 300 rem will make you sick. 1,000 rem
will kill you within days from radiation illness.
2,500 rem will induce one cancer. But how can
you ever get cancer, if the dose will kill you first?

THE PHYSICS: Biofuels are made from plants, and plants get their carbon from the carbon dioxide in the air. Since air contains radioactive carbon (C-14; see page 10), all living plant matter is radioactive. But after a few million years, all of the radioactive carbon atoms are gone; they have all radioactively exploded and turned back into nitrogen. So fossil fuels, which are made from plants buried hundreds of millions of years ago, have lost all their radiocarbon.

You can use this radioactivity as a test. If someone claims to be selling you "biofuel"—considered good for the environment—you can check if he's telling the truth by measuring the radioactivity, which is present only in authentic biofuel.

Biofuels are radioactive. Fossil fuels are not.

THE PHYSICS:

Anybody who has put wet clothes out to dry in the sun knows the amazing power of sunlight. The sun provides one gigawatt of power per square kilometer. Since the best solar cells are only 43% efficient (made by Boeing for space applications), you actually need 2.3 square kilometers of cells for that gigawatt. A square mile is 2.56 square kilometers, so it could provide 1.1 gigawatts. That's more power than a typical nuclear power plant provides.

Unfortunately, such high-efficiency solar cells are pretty expensive, and the cheaper ones are only one-third as efficient. So you might need 3 square miles of cheap solar cells for that gigawatt.

That might seem like a lot of solar cells, but they are based on a technology similar to that of computer chips, and they are getting less expensive every year. We have lots of land. Nevada by itself has 110,567 square miles, most of it usually sunny (except at night, of course).

A square mile of sunlight can produce more electric power than a large nuclear power plant.

THE PHYSICS: Spy satellites fly in low Earth orbit in order to get close enough (within 200 miles) to obtain a good view of the ground below. To keep from falling, the satellite moves around the Earth at 5 miles per second.

The satellite doesn't have to wait until it is exactly overhead to take a photo. When it is still 200 miles up-range, the total diagonal distance to its target (including altitude) is 280 miles. It continues to have a pretty good view until it is about 200 miles down-range. So it can take photos over a path of about 400 miles. At 5 miles per second, traveling that distance takes the satellite 80 seconds. If you've watched a satellite pass overhead, you know this is correct; you can see it for only a little over a minute. Since the satellite takes about 90 minutes to orbit the Earth, you might think it could then get a second view. But the Earth is rotating, so when the satellite comes back the target has moved, typically 1,000 miles away from its previous position. Twenty-four hours after its first view, however, the satellite will once again be over the target.

Weather satellites fly much higher—at an altitude of 22,000 miles—and because the gravity up there is so weak, they move so slowly that they take 24 hours to orbit the Earth. Since the Earth turns once per day, the satellite stays above the same place on the Earth. We say that these satellites are "geostationary," even though they are moving. You might think a geostationary satellite would make a good spy satellite, but they're so high that they can't resolve things smaller than about 20 feet in diameter. That's good enough for spotting thunderstorms, but not good enough to recognize Osama bin Laden.

Amateurs report the orbits of spy satellites on the Web. So Osama can check the Web and then duck under cover for a minute or two to avoid being seen.

Spying is tough. A spy satellite, in
low Earth orbit, can view a target on the Earth
for only about 80 seconds, and then it can't
return to the same target until 24 hours later.

THE PHYSICS: Light travels at 186,282 miles per second. That means in one-billionth of a second (a typical cycle time for a slow computer) light travels one foot. Electric signals go even slower. That's why computers have to be so small—so that one part of the computer has time to send a message (for example, "give me data") to another part before the next cycle tells it to move on. In other words, microprocessors, the brains of laptop computers, need to be small so that they can think fast.

HAL, the computer in the movie *2001: A Space Odyssey*, was big—sizable enough to fill a large room. When, at the end of the movie (spoiler alert!), Dave dismantles HAL's central brain, we see that it is divided into a whole series of pieces spanning several feet.

It is possible that HAL got around the size limit by using a distributed system with many separate microprocessors working in parallel. But it's more likely that back in 1968, when the movie was made, even the great futurist Arthur C. Clarke (the movie's co-writer) didn't appreciate the rapid approach of the light-speed limit—the fact that computer clock rates were increasing so rapidly that computers would soon have to be made small.

HAL, the computer from *2001*, was
way too large to be that smart.

THE PHYSICS: Many people know that liquid hydrogen has more energy per pound than gasoline—in fact, it has 2.6 times as much. But hydrogen is so light that a pound of gasoline takes 11 times as much space—that is, 11 times as many gallons. Put those two numbers together and you find that, for the same energy, liquid hydrogen takes up 4 times as much room for the same energy. Compressed hydrogen gas takes up even more space, about twice that of liquid hydrogen.

So even though hydrogen is a clean fuel (no carbon dioxide in the exhaust) and is a great fuel per pound, it is a poor fuel per gallon.

Yet light weight does have advantages for rockets and aircraft. Liquid hydrogen is the fuel for the main engine of the space shuttle. And a new high-flying spy aircraft, the Global Observer, also uses liquid hydrogen for fuel.

Want a car that uses liquid hydrogen for fuel?
Your gas tank would have to be 4 times larger
than it is now, not counting the large
thermos to keep it at −423°F.

THE PHYSICS: In 1995, the U.S. government released its previously classified Roswell files. They showed that the Roswell incident was the crash of a giant system that was part of Project Mogul. The purpose of the project was to detect and measure Russian nuclear tests. The flying part consisted of a long string of balloons that carried radar reflectors, radio transmitters, and a string of disk microphones—the most sensitive kind at that time (you can see them in old movies). The microphones listened for the rumbling of a distant mushroom cloud. It worked. The Mogul system detected the first Russian test in 1949.

In one of the earlier flights, in 1947, the flying microphone system crashed near Roswell, New Mexico. A spokesman first announced that the government's "flying disks" (the nickname used by the scientists for the disk microphones) had crashed; then he realized he had released classified information, so he tried to recall the press release. He issued a new one claiming it was only a weather balloon. Too late; people had witnessed the big pile of debris and knew it was too large to be a weather balloon. Newspapers used the earlier release, misinterpreted the words "flying disks" to be "flying saucers"—alien spaceships—and a legend was born.

When flying disks crashed near Roswell, New Mexico, in 1947, the U.S. government lied. It claimed that only a weather balloon had crashed. The government now admits that flying disks did indeed crash, and that they were recovered.

THE PHYSICS: Look really close at a TV or computer screen that is displaying the color white and all you'll see are red, green, and blue dots—no white ones. (You may need a magnifying glass to see the tiny individual dots.) But if you sit back, they blur together and make white.

Your eye has three color sensors: one each for red, green, and blue (RGB). True white light from the sun contains all the colors of the rainbow, not just RGB. True white makes all three RGB sensors respond.

Here's the trick: When the screen wants to deceive your eye into thinking the color is white, it lights up all three colors. That excites all three RGB sensors in your eye, in exactly the same way they would be excited by true white sunlight. Since that's all your brain knows—that all three sensors have detected light—it interprets the stimulus as white light. Your brain is fooled.

Some people are reputed to have four sensors, not three, so they would not be fooled by false TV white. What would it look like to them? We don't have the words to describe the colors they see, because so few people have that ability. And those few probably would call the rest of us "color-blind" for getting TV white confused with true white.

Your color TV uses one of the greatest optical
illusions of all. It makes red, green, and
blue dots appear white.

THE PHYSICS: When an earthquake fault slips, the shaking spreads out as a wave; and when that wave reaches you, you feel the shaking. It's similar to a sound wave, but it moves in the Earth rather than in the air. The quickest-moving earthquake wave (over 1,000 miles per hour) is called the "P wave" (for "primary"). It has a forward-backward shaking. The second wave to arrive, the "S wave," shakes from side to side. If you are 5 miles from the epicenter, then the P wave takes 16 seconds to reach you, and the S wave takes 17. If you are 10 miles away, the P wave comes in 32 seconds, and the S wave in 34. So take the difference in time, multiply by 5, and that's the distance to the epicenter in miles.

As a child, I was taught that I could tell the distance to lightning by counting the seconds between the flash and the thunder. For each 5 seconds of delay, the lightning is one mile away. (And to a child, a mile seems like infinity, so it is very comforting.) The calculation for earthquakes is similar, but for each second the epicenter is 5 miles away.

Next time an earthquake hits, count the
seconds between the first and second shakings
(while diving for cover). Multiply that number
by 5 and you'll know how many miles
away the epicenter is.

THE PHYSICS: Because the whole Earth—in fact, the sun and all the nearby stars (the entire Milky Way Galaxy, and we along with it)—are moving at the same speed, we don't notice how fast our planet is speeding along. It's like being inside a high-quality railroad car, or a ferry that is about to dock but hasn't yet jiggled against the pier: it's hard to tell you're moving unless you look out the window.

Cosmic microwaves come from a distance of about 14 billion light-years, so they can be used to detect motion with respect to the *really* distant matter. In 1977, I led a team that was the first to measure this high velocity of the Earth. The team included George Smoot (who later won the Nobel Prize for a follow-up measurement) and Marc Gorenstein, and the project used a U-2 airplane to detect microwave radiation from different directions in space.

For our local speed whizzing around the axis of the Earth, see page 126.

The Earth is moving at a million miles per hour in
relation to the distant stars of the universe.
You never noticed?

THE PHYSICS: When we say that electrons "spin," we mean that they turn around their own axis, just as an ice skater spins while staying in one place on the ice. Except the electron spin is much faster.

We don't really understand why electrons spin, although we can describe the spin very accurately with equations. (Equations don't really *explain* anything; they just turn the physical mystery into a mathematical one.) We know we can change the direction of the spin but we can't stop it.

Electrons have an electric charge, and since the charge spins too, electrons are little magnets. Permanent magnets are made of materials in which many of the electrons are lined up to spin in the same direction. That's how the magnets in your earphones and on your disk drive get their strong magnetism—from a large number of spinning electrons.

Electrons spin and we can't stop them.
Nobody has even seen an electron
whose spin has stopped.

THE PHYSICS: Batteries are expensive, but very convenient. Flashlights work even when the electric grid fails, but you have to pay a lot extra for that convenience. A typical (nondiscounted) price for a AAA battery is $1.50 (that's the price at a convenience store here in Berkeley). It delivers 1.5 volts and 1 amp—multiply those together to get 1.5 watts—for about an hour. That's $1.50 for 1.5 watt-hours, or $1,000 per kilowatt-hour. If you buy that much energy from your local electricity utility, you'll pay only about 10 cents.

To most people, it's pretty obvious that battery electricity is much more expensive than wall-plug electricity, although few people realize it is 10,000 times more costly. But think about it: If you leave your home and remember you didn't turn off your flashlight, you'll probably go back to turn it off. You might not bother for a bathroom lightbulb.

A hundred and fifty years ago, virtually all electricity came from batteries. Early telegraphs used batteries for power. Some people predicted, back then, that someday electricity would be "too cheap to meter." I suspect they would have considered 10 cents per kilowatt-hour to be pretty close to that zero cost. Of course, since electric energy is now so cheap, we squander vast amounts. We even make rooms without windows and use electricity to light them during the day.

Electricity from your wall plug costs typically
10 cents per kilowatt-hour. The same amount
of electric energy from AAA batteries
would cost you $1,000.

THE PHYSICS: There was once much more uranium-235 (U-235) on the Earth than there is today, since it had not yet decayed away from its own radioactivity. At a location near Gabon, Africa, seeping water and this "enriched" uranium (see pages 118 and 120 for details about uranium enrichment) formed just the right combination to turn into a nuclear reactor, producing large levels of radioactivity and about 15 kilowatts of heat.

About 1.7 billion years later, miners discovered that in this region there was less than the usual fraction of U-235; it had been depleted, used up. That was the first evidence that there had been an ancient reactor. Subsequent investigation found the rare isotopes called "fission fragments" that are created in a reactor, and that discovery proved the theory.

Why didn't the reactor explode? Even though there was more U-235 back then, it was enriched to only 3%. (Currently, natural uranium has only 0.7% U-235.) To make a bomb requires a much higher level, near 100%. That's also why modern nuclear power reactors can't blow up like an atomic bomb: the uranium in a modern reactor is enriched to only 3%, the same level that made the Gabon reactor work.

The world's first uranium nuclear reactor is
1.7 billion years old.

THE PHYSICS: A "gigawatt" is a billion watts, enough to supply a million homes, assuming each house uses an average of one kilowatt (1,000 watts). There are a million grams to a ton, so at the level of coal use specified in the illustration, each home is using only one gram of coal every 7 seconds. That doesn't sound like so much, but for a million homes that adds up. No wonder power plants get so huge!

When the coal is burned, each atom of carbon combines with two atoms of oxygen to produce one molecule of carbon dioxide. In other words, C combines with O_2 to become a molecule of CO_2. (The "di" in "dioxide" means "two," so carbon dioxide has two oxygens, represented by O_2.) The CO_2 molecule weighs a bit more than 3 times as much as the carbon by itself. The extra weight comes from the oxygen taken from the air. That's why so much more carbon dioxide (in tons) is produced than coal is burned.

A gigawatt (big) coal-burning power plant burns
one ton of coal every 7 seconds. And every
2 seconds it releases one ton of carbon dioxide,
the notorious greenhouse gas.

THE PHYSICS: We fell in love with gasoline not because of its physical beauty, but because of its personality. Gasoline delivers an enormous amount of energy—15 times as much as an equal weight of TNT. And in the past, gasoline seemed to be not only cheap but also clean, since its primary emission was good clean carbon dioxide, the gas that humans breathe out and plants need to grow! Compare gasoline to coal, which not only emits a lot of soot, sulfur dioxide, and mercury, but leaves behind a pile of ash. In contrast, when you've burned the gasoline in your car, your tank is simply empty. Nothing to clean out. Wow.

And gasoline is safe! After all, the only downside might be war in the Middle East, and that was highly unlikely, since the West owned all the oil resources.

Things have changed. Carbon dioxide is now recognized as a potentially dangerous greenhouse gas. And we have lost thousands of U.S. soldiers in oil-region wars.

The problem is that we are not really having a love affair with gasoline; we are married to it. We want a divorce, but we have grown so dependent on it that any divorce is bound to be ugly.

Why do we love gasoline so much?
It certainly isn't the smell!

THE PHYSICS: Meteors typically go very fast, about 20 miles per *second*. That's 20 times faster than a bullet from a high-powered rifle. The energy of motion depends on the square of the velocity (20 × 20), so the meteor has 400 times the energy of the bullet, for the same weight.

If the meteor is big enough to be called an "asteroid," then the Earth's atmosphere doesn't slow it down much. When it hits, all its energy is turned into heat, and that heat vaporizes rock, making an explosion 100 times bigger than you would get from exploding an equal weight of TNT. That's what happened 66 million years ago, when an asteroid the size of San Francisco hit the Earth and killed the dinosaurs and almost everything else alive at that time. The impact turned the energy of motion into heat, and that heat vaporized the asteroid. Hot gases rushed out, causing the explosion that destroyed so much life.

Here's a look at the numbers: Physicists calculate kinetic energy from the equation $KE = \frac{1}{2}mv^2$. The hard part is using the right units. Mass must be in kilograms, and velocity must be in meters per second. If we have a mass m of 0.5 kilogram moving at a velocity v of 30,000 meters per second, the kinetic energy KE is 225 million joules. The energy released by a pound of TNT is about one kilocalorie per gram, or 1.9 million joules per pound—less than 1% of the energy in a meteor.

A one-pound meteor heading toward the Earth carries 100 times more energy than one pound of TNT.

THE PHYSICS: The crust of the Earth contains uranium and thorium, and both of these elements are radioactive. The energy from this radioactivity is the source of the heat for volcanoes, geysers, hot springs, and geothermal energy plants. So geothermal energy is really nuclear energy.

When the uranium and thorium atoms explode, they emit radiation called "alpha rays." An alpha ray is really an energetic *particle* consisting of two protons and two neutrons stuck to each other. When an alpha particle slows down, it attracts two electrons to orbit it, and the result is an atom of helium. This gas tends to accumulate in oil pockets, and when we drill for oil, the helium gas comes out too. That's where we get our helium for party balloons.

The children in the illustration don't realize that not all nuclear waste is dangerous. In fact, the helium—the waste from decay of uranium and thorium—is not itself radioactive. No need to worry about helium balloons!

The helium we put in children's balloons is
actually waste from radioactivity.

THE PHYSICS: The excess radioactivity near Three Mile Island did not come from the accident, but from natural radon seeping out of the ground. Radon is a decay product of the relatively high natural uranium levels found in this region. (Uranium and thorium also produce helium gas, but as discussed on page 62, it isn't radioactive.)

After the Three Mile Island accident, in which a tiny fraction of the reactor radioactivity was released, people began for the first time making extensive Geiger counter measurements in the vicinity. At first, they were horrified to find levels 50% greater than they had expected, and even though such levels are still safe, this discovery caused some degree of panic and concern. But then they identified the radioactivity as coming from radon and found it was seeping out of the ground. The radon had been coming out for millions of years but had never been noticed.

Other locations around the United States have similar or even higher radon levels (see the question about evacuating Denver, on page 68). Even some buildings made from granite show high radon levels, since granite tends to have high levels of uranium. ("High" in this case means parts per million.) The U.S. government advises people living in high-radon areas to measure the radon levels periodically in their homes. If they find dangerous levels, the standard treatment is easy: increase air circulation so that the radon quickly leaves the house.

By the way, 50% greater radiation would not have produced the dinosaur mutations seen in the illustration. That's a bit of an exaggeration. For more on Three Mile Island, see page 66.

The amount of radioactivity in the region surrounding Three Mile Island in Pennsylvania, the site of the 1979 reactor accident, is 50% higher than the average level in the United States. But it has been that way for millions of years.

THE PHYSICS: The worst nuclear power plant accident in the United States occurred in 1979, when part of the core of a nuclear reactor at Three Mile Island in Pennsylvania overheated and melted, and radioactivity was released into the air.

The official Kemeny Commission studied the harmful consequences of the incident. The commission concluded that the radiation exposure to the public was only 2,000 person-rem, spread out over millions of people. No one person got a dose greater than 0.07 rem, which is well below the 0.40-rem exposure that people get every year from natural radioactivity and X-rays.

But, the commission stated, "The major health effect of the accident appears to have been on the mental health of the people living in the region of Three Mile Island and of the workers at TMI." Severe distress was caused by the mistaken belief that the disaster was gravely threatening the lives and homes of the people who lived nearby.

Distress often leads to extra smoking, and that leads to lung cancer—but the Kemeny Commission didn't try to evaluate that particular danger.

The notorious nuclear accident at
Three Mile Island did hurt people—but not
from the released radioactivity. It was panic
that caused the most harm.

THE PHYSICS: The high level of radioactivity in Denver is due to the traces of uranium in the Rocky Mountains. The extra dose for people living in the Denver area is about 0.1 rem per year. That's the equivalent (whole-body dose) of about three chest X-rays, or thousands of tooth X-rays. By the linear hypothesis, it should give residents an extra chance of cancer of about one part in 25,000. That's a small risk, but the population of Denver is about 600,000, so the extra radiation should trigger about 24 additional cases of cancers each year.

Yet the cancer rate is actually lower in Denver than in the rest of the country. Some scientists argue that this means small levels of radiation are good for you—that weak radiation "inoculates" you, protecting you from higher levels. But it is also possible that other factors (for example, lifestyle) are the real reason for lower cancer levels in the Denver region. Such cancer-lowering factors could be obscuring and canceling a radioactivity-induced increase; or maybe low radiation levels really are good for you. Or not. We don't know.

The level of radioactivity is 50% higher in Denver than in most of the rest of the country. If this high level were the result of a recent occurrence, Denver would have to be evacuated.

THE PHYSICS: A "storm surge" is an enormously high tide that brings the ocean right into your living room. Part of it comes from low atmospheric pressure that sucks up the ocean. The surge is higher on the side of the storm where the winds push even more water toward the shore. Storm surges are particularly bad if they happen during a normal high tide. The storm surge in Hurricane Katrina was recorded as reaching 25 feet above normal sea level in some places. The Outer Banks of North Carolina has islands that become completely submerged in storm surges from large hurricanes.

Winds can do a lot of damage too—but water is almost a thousand times more dense than air, so even if it is moving a lot slower, it can exert a much larger force.

Most deaths in a hurricane are not caused by the winds, at least not directly. They are caused by the "storm surge," the rise of water that submerges coastal towns under ocean waters.

THE PHYSICS: The altitude of the International Space Station is about 200 miles. At that altitude, the force of gravity is only 10% weaker than on the Earth's surface. So a 150-pound astronaut standing on a stationary scale would weigh 135 pounds.

Yet, if an orbiting astronaut stood on a scale, the scale would read zero. Why does the astronaut seem weightless in this case? The reason is that the scale is orbiting the Earth along with the astronaut. It's as if they were all in a falling elevator. If everything falls together, no force is exerted by the feet on the scale. If you are in orbit along with the scale, you feel no force on your feet or in your leg muscles, so you feel weightless even though you aren't.

If the gravity were truly zero, the space station and the astronauts in it would just fly off in a straight line, not curve downward and around the Earth. The velocity for satellites is chosen so that their downward curving just matches the curvature of the Earth and they stay in a circular orbit. In fact, you can think of an astronaut as constantly falling but always missing the Earth. Being in orbit is like being infatuated: you are constantly falling, but you aren't getting closer.

**"NO, THIS DOESN'T COUNT AS
A WEIGHT-WATCHERS RECORD."**

Stand on a scale in space, and it will read
"0 lb." Yet gravity in orbit is almost as strong as
on the Earth's surface. So why are
astronauts weightless?

THE PHYSICS: Going up 100 kilometers isn't that hard. Germans sent a V-2 rocket up 85 kilometers back in 1942. In 1946, the United States reached 185 kilometers (116 miles) by sticking a small rocket (called a WAC Corporal) on top of a captured German V-2. But it took another decade, a rocket race between superpowers, and a lot more money before the first satellite was orbited by a Russian rocket in 1957.

The trick isn't getting up there; it's *staying* up. To do that takes an extremely high velocity, about 5 miles per second, so that you stay in a circular orbit. That feat takes about 30 times more energy. A factor of 30 can be more dramatically expressed as 3,000%.

That means that the X Prize rocket was 30 times too weak to send its astronauts into even low Earth orbit. So why was the X Prize success so important? Technologically, it wasn't. But it tended to excite people about space travel, and that made some people (especially in NASA) happy.

The X Prize—winning rocket sent a man 100
kilometers high. He was then proclaimed to be
an astronaut. How much more energy would
have been needed to send him into orbit? The
surprising answer: 3,000% more.

THE PHYSICS: Not all molecules move at the same speed, even when they're at the same temperature. The physics law is that light and heavy molecules all have the same energy, on average—not the same velocity. That means light ones have much higher speed. A typical hydrogen molecule moves 3 times faster than a typical oxygen molecule.

That's the average speed; some go faster, some go slower. The faster hydrogen molecules actually get up to the speed of escape velocity, leaving the Earth like a rocket. It took a while, many millions of years, but eventually all the hydrogen molecules in the atmosphere escaped in this manner. After 4.6 billion years (the age of the Earth), there is very little left. The only hydrogen atoms left behind were those that had combined with other elements in more complex molecules, like water.

The big planets in the solar system, such as Jupiter and Saturn, are harder for gases to escape from because of their stronger gravity. So they still have lots of hydrogen gas in their atmospheres, just as the Earth once did.

"I HEARD THEY COULDN'T LEAVE FAST ENOUGH."

Hydrogen gas molecules move faster than those of any other gas. That's why there's almost none left in the atmosphere. They escaped to space.

THE PHYSICS:

Certain "isotopes" (atoms with an unusual number of neutrons in the nucleus) are radioactive; and when they decay, some of them emit positrons—the antimatter form of electrons. Positrons have proven enormously useful for diagnosis, in the device known as the PET scanner. (PET stands for "positron emission tomography.")

One of these positron emitters is iodine-124. (That's like ordinary iodine, but with three fewer neutrons in the nucleus.) If iodine-124 is fed to a patient or injected into a vein, it tends to be absorbed by the thyroid gland. It has a half-life of only 4 days, so the iodine soon emits antimatter positrons into the thyroid. These positrons go only a very short distance before colliding with ordinary electrons. On coming into contact, positrons and electrons "annihilate" each other: both positron and electron disappear, and they release all their energy in a pair of gamma rays. These gamma rays are easily detected outside the body, and the PET technician can tell where they're coming from.

One useful application of PET is to locate regions of the thyroid that are dead (no iodine goes there, so there are no gamma rays from that part) or are hyperactive (absorbing more than their fair share of the iodine, and creating more gamma rays).

PET scan measurements are made every day in large hospitals. Antimatter has become a daily and very useful tool for medicine.

Antimatter is not science fiction fantasy.
It is used in virtually every large hospital in the
United States to help diagnose cancer
and other illnesses.

THE PHYSICS: A white surface is one that reflects all the wavelengths of sunlight and contains all the colors of the spectrum. But some of these colors are invisible to the eye. Suppose we added a chemical, a "phosphor," that converted some of the invisible ultraviolet light into additional visible light. Then clothing would look extra bright when examined in sunlight—brighter than clothing that only *reflected* all of the light and left the ultraviolet invisible.

People who make white clothing add phosphors to make the clothing appear extra bright. Laundry detergent manufacturers also add phosphors to their mix. Some phosphor sticks to the clothing; when people see the brightness, they mistake it for extra cleanliness. In fact, since this trick works for laundry detergents only if the phosphor remains behind, it isn't true that the brightness means the clothes are actually cleaner. They are dirty with phosphor.

Phosphors embedded in the fibers of clothing are what make some white shirts light up under black light (used in some amusement parks and in Halloween displays). The black light is mostly ultraviolet, and it stimulates the phosphors to emit a bluish white light.

Clothes that are "whiter than white" really are.
The claim is not just an advertising slogan, but is
solidly based in physics.

THE PHYSICS: With an ordinary lightbulb, 84% of the light emitted from the hot tungsten filament is invisible heat radiation, also called infrared or IR. Lightbulbs are more efficient (that is, made to emit a larger fraction of their energy in the form of visible light) if the filament gets hotter; it's that high temperature that makes halogen lamps both brighter and more dangerous. Moreover, the hotter the filament, the more rapidly the bulb "burns out." In truth, the filament isn't burning, but the metal in the filament actually evaporates at high temperature, and when enough metal has left, the filament breaks.

Ironically, "long-life" tungsten bulbs sold in the store get their long life by operating at a lower temperature. That means their efficiency is even worse. Over 90% of the light emitted by a long-life bulb is invisible (and wasted, at least for the human eye).

We can get much higher efficiency if we generate light without using heat. Fluorescent lights give the same amount of visible light but use only one-third as much power. Light-emitting diodes (LEDs) are similarly efficient, and are now widely used in flashlights.

84% of the light emitted by a
household incandescent lightbulb is
invisible to the human eye.

THE PHYSICS: The core of the Earth is over 5,000°F—about half the temperature of the surface of the sun. But rock is a good insulator; it keeps us from frying. Strip away the rock and the heat radiation from the core will vaporize you.

The downside of the good insulating property of rock is that it makes it difficult to exploit geothermal energy (see page 86).

If you peeled back the outer layers of the Earth
and looked at the core, you would be burned to
a crisp in a fraction of a second. It's that hot.

THE PHYSICS: "Geothermal power," the heat flow that rises up from radioactivity in the Earth's rock, is responsible for volcanoes, geysers, and earthquakes. This power from the interior of the Earth is huge—about 30 terawatts. That's a 3 followed by 13 zeros. Sounds big.

But the surface of the Earth is big too, about 500 trillion square meters. That means the average energy flow is only 0.06 watt per square meter. That's about 1/10,000 of the energy flow from sunlight (if you average in nights as well as days). This means that to collect enough geothermal power to make it competitive with solar power, it would have to cost 1/10,000 as much per square meter. This constraint makes geothermal energy impractical over most of the Earth. The exceptions are places (such as Iceland) where underground flow of heat and magma concentrates the power into a small region.

Geothermal power is huge, but so
spread out that it isn't practical to tap unless
the Earth cooperates.

THE PHYSICS: Diamonds were once valued because of the way they sparkled. They have very high "dispersion," meaning that white light is strongly broken up into the colors of the rainbow. A diamond that is cut in a complex way sends out many different colors in different directions.

Of course, diamonds were once rare, and that made them expensive, so you would buy one as a present only for someone you truly loved (or wanted to impress).

Now we can make artificial crystals known as cubic zirconia that have a higher dispersion than do diamonds and are much cheaper. So if you give one to your loved one, what does it prove? Good taste? Intelligence? Okay—but does it prove love, if it is cheap?

Many people apparently think the price is more important than the beauty. So diamond jewelry is no longer expensive because it's beautiful; it's expensive because the diamonds cost a lot. Diamonds are not really rare either, except to the extent that the diamond cartels try to keep them off the market to keep the price high.

Another problem with cubic zirconia is that it's so much prettier than diamond that it's easy to tell it isn't diamond. Physicists are working hard to reduce the beauty of cubic zirconia stones so they will look more like the expensive ones.

Phony diamonds, made of "cubic zirconia,"
are prettier than real diamonds. Their main
disadvantage is that they cost less.

THE PHYSICS: Just as the word "transistor" is short for "transistor radio receiver," and "microwave" is short for "microwave oven," the term "speed of light" as used by physicists is short for "speed of light in a vacuum." When physicists say that "nothing can go faster than the speed of light," they mean that nothing can exceed the speed that light travels in a vacuum—186,282 miles per second.

Light slows down when it travels through transparent materials such as water or glass. That's because its electric field makes the atoms shake; it's as if the light were acquiring mass. The amount it slows down is called the "index of refraction." For water, the index of refraction is about 1.33, and for window glass it is 1.5. For diamond, the index is 2.4. That means that the speed of light in a diamond it is only 77,618 miles per second. Lots of things travel faster than that; well, at least subatomic particles accelerated in human-made atom smashers do. In principle, you could too.

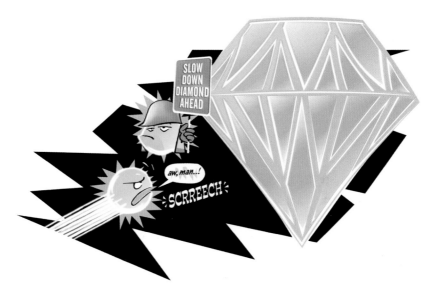

Light doesn't always travel at the speed of light. It goes slower through water and glass and diamonds. So yes, you can travel faster than the speed of light, if what you mean is faster than the speed of light *in diamonds*.

THE PHYSICS: The ozone layer is created by the UV (ultraviolet) light that comes from the sun. This energetic radiation breaks up O_2 molecules; some of the free oxygen atoms stick to other O_2 molecules and make O_3—"ozone." By blocking most of the rest of the UV light from reaching the surface of the Earth, ozone protects us from sunburn and skin cancer. But this UV blocking absorbs energy and heats up the upper atmosphere in a region called the "ozone layer."

As you climb to higher altitude, the temperature of the air initially gets cooler. That's why mountaintops often retain snow in summer. It's cold up there because rising air (from wind) expands, and when it does it cools. But if you get up to the stratosphere, where UV is absorbed by ozone, the atmosphere gets warmer again.

Thunderheads contain warm, light air, and they rise through cool, heavy air. They basically float up through it, like a bubble in water. But if they reach the warm air at the ozone layer, they don't rise any more. They spread out, making the notorious "anvil head" cloud that forebodes a severe storm.

So to locate the ozone layer, look for the anvil head.

The ozone layer—the part of the atmosphere that protects us from dangerous UV sunlight—is easily spotted when a storm is approaching on a hot summer day. Just look up at the sky and note where the thunderhead stops growing and begins to spread.

THE PHYSICS: Anyone who has flown a kite knows that the strongest wind is up high. That's because friction with the ground slows down the low air currents. So to reach the best winds, modern wind turbines are built tall.

The Statue of Liberty is not as tall as many people think— 151 feet from the base to the tip of the torch. That's about the height of a 15-story building—small by New York City standards. (It's big in meaning, though, and it's large compared to anything else nearby.) In contrast, the Enercon wind turbine in Germany has a hub height of 453 feet—and its blades reach up another 150 feet—for a total height of about 600 feet!

Some modern wind turbines are 4 times as high
as the Statue of Liberty! Why?

THE PHYSICS: In 1903, Thomas Edison was in the midst of a dispute with Nikola Tesla. Edison favored low-voltage DC for electrifying cities, with a coal-burning electric power plant every few city blocks. Tesla (and his supporter George Westinghouse) favored high-voltage AC electricity, which enabled transmission lines to be more efficient so that the power plants could be located far away.

Edison argued that high voltage was dangerous. (You may want to stop reading here.) To illustrate, he made movies showing how high voltage could kill small animals such as puppies, and then larger animals, including a horse. When a circus elephant named Topsy killed three men (including one who had fed her a lighted cigarette), she was condemned to die. Edison claimed that electrocution was effective and humane, and the American Society for the Prevention of Cruelty to Animals (ASPCA) approved. Topsy was electrocuted (the tragedy became an inspiration for the movie *Dumbo*). Eventually, Edison convinced New York State to use electrocution for executing condemned humans.

Even so, Tesla won the battle for high-voltage AC, and that's what we use today. The voltage is reduced by transformers before it enters the home. Transformers don't work with DC.

AC stands for "alternating current," meaning that the electricity flows back and forth, making 60 full cycles every second (50 in Europe). DC stands for "direct current," meaning it always flows in the same direction. Batteries give DC.

Thomas Edison suggested using electricity to
execute an elephant. Topsy, a "bad elephant,"
was killed in this way. Edison filmed the event,
and the movie can still be found on the Internet.
I don't recommend watching it.

THE PHYSICS: The magnetic compass was once a closely guarded military secret. A magnetized needle, originally floating on water, oriented itself to the magnetism of the Earth. It was invaluable for military ships navigating on cloudy and stormy days and nights. And commercial ships too.

The first clear use for the magnetic compass for sea navigation goes back to the 1100s in China and in Europe. But as with GPS, the modern tool for navigation, the magnetic compass was so valuable that the secret quickly got out and its use spread rapidly. In 1620, Francis Bacon ranked the compass as one of three inventions that had revolutionized the world. The other two were gunpowder and the printing press; he must have meant "*recent* world," since he didn't include the wheel.

The magnetic needle of a compass feels the force from the magnetic field of the Earth. The fact that the Earth was a magnet wasn't known until much later, the 1600s. Before that, most people thought the needle was attracted to the North Star. But William Gilbert, the personal physician of Queen Elizabeth I, wrote (in Latin), "A magnificent magnet is the terrestrial globe."

For more on the Earth's magnetism, see page 100.

Magnetism was once a military secret.

THE PHYSICS: Try to follow this logic: the north-pointing end of a compass is attracted to the south pole of a magnet. But it is also attracted to the geographic North Pole. That means that the geographic North Pole is actually a south magnetic pole.

But it wasn't always. One of the most amazing discoveries ever about the Earth is that the magnetism of the our planet changes, at somewhat random times. When the switch is about to take place, first the magnetism turns off—that takes a few hundred years. We then have a few thousand years with virtually no magnetism, after which, it turns back on, often in the opposite direction. When that happens, it's called a "magnetic flip." If it turns back on in the original direction, the incident is called a "magnetic excursion."

The last time a flip happened was about 800,000 years ago. We don't understand why, although there are many theories. (I published my own theory in the peer-reviewed scientific literature, but it hasn't yet been shown to be either right or wrong.) No matter what the explanation is, these flips are recorded in sedimentary rock and are enormously useful for estimating the age of such rock.

The North Pole is actually a south pole,
magnetically speaking. But 800,000 years ago
the North Pole really was a north pole.

THE PHYSICS: According to three of the four major science teams that measure global temperatures, the warmest year on record was 1998. Temperatures since then have been lower. But that does not necessarily imply that global warming has stopped. The halt in warming may be due to a temporary reduction in solar activity (sunspots have been very weak recently) or caused by a drying out of the upper atmosphere (which has been observed).

So don't get complacent. Carbon dioxide continues to increase, and unless something else cancels it, the physics of greenhouse warming is pretty straightforward: increased carbon dioxide reflects more of the Earth's heat radiation back to the surface, and that should cause warming. But the atmosphere is complex, so we can't be sure. The UN commission that studies this says such warming is 90% certain.

Eventually—maybe in 100 years or so—the increased carbon dioxide will start making the oceans significantly more acidic. So even if global warming stops, we'll still have something serious to worry about.

There has been no significant global warming
over the past 12 years.

THE PHYSICS: Light travels at 186,282 miles per second, but the sun is 96 million miles away, so it takes light about 8.6 minutes to reach the Earth. No explosion, no signal of any kind, can outrun the speed of light in a vacuum. So anything that happens on the sun, even a supernova explosion, would not reach us for at least 8.6 minutes. We wouldn't know it had happened until 8.6 minutes after the fact.

Likewise, when you look at a star, you're not seeing it the way it is right now. Sirius, the brightest star in the sky, is 8.6 light-years away. So when you look at it at night, you're seeing it the way it was 8.6 years ago. Maybe it's not even there anymore!

If the sun blew up 8 minutes ago, you wouldn't
know it yet. But just wait another minute.

THE PHYSICS: Although Hurricane Katrina was a "category 5" storm (the worst kind) while out at sea, it was only a category 3 storm when it hit New Orleans. Category 3 storms are common. Any category 3 storm in the last 30 years that hit New Orleans could have destroyed that city. But they all missed. Until Katrina.

Many people mistakenly think that the number of hurricanes, especially the violent ones, are increasing because of global warming. But according to the U.S. National Hurricane Center, that's not true. We just observe more because we're looking harder.

In the old days, we knew the strength of a storm only when it actually hit the shore. Then, in 1943, the United States established the "Hurricane Hunters," brave pilots who flew out to sea seeking hurricanes whose existence was suggested by waves and reports from ships. In the 1960s, we began to make wind velocity measurements by satellite. Buoys were also put out at sea to detect waves and surface wind speeds. In 2005, the United States started sending out unmanned aircraft called drones.

Largely because of such observations, we detect many more hurricanes today than we did 50 years ago. In 2005, after Hurricane Zeta, we had to start reusing letters of the alphabet for hurricane names for the first time. Hurricanes don't seem to be increasing in number (if you look at the number that hit the U.S. coast every decade, it is constant), but our detections are going up.

Some people think that average hurricane winds are also getting stronger, but that, too, is a misperception. Any hurricane that even briefly qualifies as a category 5 storm (wind speeds of 156 miles per hour and up) is identified that way by the newspapers even after its strength diminishes. Katrina was category 5 while out to sea, but it slowed to category 3 (winds 111–130 miles per hour) before it hit New Orleans. It was the first category 3 storm to strike New Orleans since the levees had been built.

Hurricane Katrina was not an unusually large
storm when it hit New Orleans. There have
been dozens of similar storms over the past
several decades, and any one of them
could have drowned that city. Katrina was just
the first one that happened to hit.

THE PHYSICS: We really are made of ashes. How do we know? The Big Bang was so hot that the only elements that survived were hydrogen and helium (okay, maybe a little bit of lithium, according to some theories). But we are made of carbon and oxygen, as well as hydrogen. (Those are the constituents of carbohydrates.) If they weren't made in the Big Bang, where did our carbon and oxygen—the elements necessary for life as we know it—come from?

They were created in the cores of stars, through the process known as "thermonuclear fusion." In one of the reactions, three helium nuclei come together to make carbon, and oxygen is then made from the carbon.

If these elements had remained trapped inside a star, life still would not be possible. But large stars sometimes explode in what is called a "supernova," an event that spews the elements out into space, where they sit until some other cosmic event occurs. It was in such a debris-filled region that the mutual gravity of the supernova debris drew it together into a dense ball that would eventually be called the "Sun." Some of the debris was left behind orbiting the sun; this residue eventually clumped to form planets, including the Earth, full of the elements of life. And here we are.

Humans are the debris of an exploded star.

THE PHYSICS: To grow organic foods, farmers must select plants that are "naturally resistant." How do these plants get that way? The great biochemist Bruce Ames investigated and discovered that plants develop resistance by creating poisons that kill the insects and other creatures that might try to eat them. He says, "Plants can't run—so they use chemical warfare."

Ordinary farmers choose any plants they want—including plants that have poor resistance to insects—and then protect by spraying them with artificial pesticides. In contrast, organic farmers choose plants that are *naturally* resistant—in other words, plants that have skin and flesh *naturally* full of poisons. The low level of these poisons is not enough to kill humans, but they can induce cancer. Typically, the natural pesticides in organic food are thousands of times more carcinogenic than the artificial pesticides approved by the U.S. Department of Agriculture. And since they are part of the food, they can't be rinsed off.

Incidentally, I eat and love organic food, but mostly because the people who grow it somehow make it taste better. Damn the carcinogens—full flavor ahead!

Organic foods are higher in poisons and
carcinogens than foods grown using pesticides.

THE PHYSICS: We have pretty good measurements of the mass of the universe, mostly from its gravity and from the rate of expansion (the Big Bang). Here's the tally sheet:

- Ordinary matter (mostly hydrogen and helium, with a little bit of heavier elements such as carbon, oxygen, and silicon) accounts for 4%. This is the stuff of stars, rocky planets, comets, and humans.

- "Dark matter" accounts for 30%. It was named that because it doesn't shine like stars. We know of it only because its gravity makes galaxies move in unexpected ways. We don't know what it is. Two leading candidates are WIMPs ("weakly interacting massive particles") and MACHOs ("massive compact halo objects"). These are the cleverest acronyms in all of physics.

- The remainder of the universe is made of something even more mysterious, which we call "dark energy." We don't even have any good candidates for what it is. We know it's there, because it's making the explosion of the universe, the Big Bang, speed up. That's why we describe it as "energy."

What is the universe made of? We don't know.
Only 4% is made of matter, ordinary atoms, and
molecules and ions, the material of stars and
planets. The rest is made of two mysterious
substances, material we don't understand, called
"dark matter" and "dark energy."

THE PHYSICS: The high cost of space travel is due to the inefficiency of rockets. Unlike airplanes, which push against air to fly, rockets get their thrust by pushing exploded fuel backward. And chemical fuel at its highest energy per gram shoots backward at less than one mile per second, much slower than the 5 miles per second needed to maintain a low Earth orbit. So acceleration is slow. The space shuttle must use 30 pounds of fuel for every pound of satellite that goes into orbit.

To make reaching space more efficient, NASA is trying to develop an airplane that could go fast enough, but such hypersonic speeds are probably decades away. Some people think a giant elevator would be the least wasteful way to reach space, but to stay in low Earth orbit you still have to get the satellite moving 5 miles per second, and currently, that feat still takes rockets.

An invention called a "rail gun" is much more efficient than a rocket. But even a mile-long rail gun pointed to space would have to accelerate the spaceship by 2,000 g's, enough to crush not only any astronaut, but most satellites. (The force put on the astronaut by 2,000 g's is 2,000 times the astronaut's weight. That amounts to one ton for every pound.) In contrast, the maximum acceleration in the space shuttle is only about 3 g's.

To launch one pound of *anything* into orbit costs $5,000 and takes 30 pounds of fuel. No wonder space hasn't been much utilized by manufacturers!

THE PHYSICS: To keep a chain reaction going in plutonium, you need a "critical mass." That's enough material that a neutron produced in a fission is unlikely to leak out but will instead hit another nucleus and trigger another fission. In just 60 such stages, you explode a large fraction of the atoms in the chunk.

The critical mass for plutonium is just 13 pounds. Because plutonium is so dense (almost twice that of lead), those 13 pounds would fit in a 12–fluid ounce soft-drink can—or a large coffee mug.

If every atom in those 13 pounds exploded, the energy released would be that of 100 tons of TNT. The plutonium in the first atomic bomb—tested at Alamogordo, New Mexico, in 1944, blew apart when only 20% of its atoms had fissioned. So its energy was *only* 20 kilotons of TNT equivalent.

MANHATTAN BLEND

"THAT'S ONE POWERFUL CUP"

Fill a coffee mug with plutonium and you'll have
enough to make an atomic bomb.

THE PHYSICS: The design of a uranium atomic bomb is quite simple: take two pieces of virtually pure U-235 ("light uranium"), and shoot one at the other to make a critical mass. For the bomb that destroyed Hiroshima, about 100 pounds of pure U-235 was used. So all a teenager has to do is obtain the enriched uranium and a medium-sized cannon.

Making a bomb out of plutonium is much harder, since it requires an implosion. That's beyond the means of a teenager.

For a uranium bomb, the hard part is getting the enriched uranium. Ordinary uranium, mostly heavy U-238, won't work for an atomic bomb. The U-238 absorbs too many neutrons without fissioning. You have to purify the uranium so that the fissionable U-235 is increased from less than 0.7% (in natural uranium) up to nearly 100%. That process, called "enrichment," is extremely difficult. It took the United States most of a year to do it during World War II, using a calutron to enrich (for more about this, see page 120). After the war, the United States used thermal diffusion through Teflon. Now, the modern technique (used by Pakistan, Iran, and others) is to use a centrifuge. All of these technologies are exceedingly difficult, well beyond the capability of a teenager.

But if the student got some rogue U-235 smuggled out of Ukraine . . .

Could a teenager make an atomic bomb?
Maybe. The hard part is getting the
enriched uranium.

THE PHYSICS: During World War II, the great physicist Ernest Lawrence came up with a way to enrich uranium—that is, to separate out the light uranium, U-235, from the much more common but useless heavy uranium, U-238. His method was to vaporize the uranium, accelerate it, and bend its path in a magnetic field. The lighter U-235 curved more in the field and came out at a different spot than the heavier U-238. By the end of the war, this machine was able to produce enough uranium for one bomb, the one dropped on Hiroshima. The other two big bombs of that era—the one tested in Alamogordo and the one dropped on Nagasaki—were plutonium bombs, not uranium bombs.

Because the uranium vapor traveled in an arc, the uranium-enriching machine was shaped like a big letter "C." That was also the symbol of the University of California, Berkeley, nicknamed "Cal." Lawrence, a professor at Cal, had already invented the cyclotron (and had received the Nobel Prize for it), and he decided to name this big C-shaped machine after his own university. He called it a "calutron."

The machine used to enrich the uranium that destroyed Hiroshima was named after "Cal," the nickname for the University of California, Berkeley. No student-led tour currently mentions this fact.

THE PHYSICS: Certainly nothing in physics explains why we use the word "the" with "Earth" when we don't do the same with other planets, and there's probably no good semantic explanation either. It's not as if there were a whole bunch of Earths needing to be distinguished. English is just strange sometimes. We might similarly ask why we say "the Bronx" but not "the Manhattan."

When we refer to the Earth, why do we say "*the* Earth" rather than just Earth? After all, we don't say "the Mars" or "the Jupiter."

THE PHYSICS: The word "planet" means "wanderer" in Greek. The ancients called star-like objects in the sky that tended to move between the "fixed" stars "wanderers." When Pluto was discovered in 1930, it was recognized as a star-like wanderer and immediately dubbed the "ninth planet."

Because we now know that Pluto is small and made mostly of rock and ice, it seems more like a comet than a planet. So the IAU had a vote at its 2006 meeting and decided it should no longer be called a planet, but only a "dwarf planet."

I say that astronomers have every right to call Pluto something different when they talk among themselves. But the IAU didn't give Pluto the designation "planet" when it was discovered in 1930; that name was awarded by the discoverer and by consensus. So what gives the IAU the modern right to demote Pluto? The answer is, nothing. Self-proclaimed authority is not necessarily legitimate authority. (I say that even though I am a member of the IAU.)

I took a poll in the class I teach, and the students reinstated Pluto as a planet by a vote of 512 to 0. Since this class is bigger than the group at the meeting of the IAU that voted to demote it, it is my professional opinion that Pluto is still a planet.

Is Pluto no longer a planet? Many astronomers think it isn't, since the International Astronomical Union (IAU) voted to demote it in 2006. But I disagree. Pluto has been a planet for 4.6 billion years, and it still is one.

THE PHYSICS: The Earth rotates about its own north–south axis once every 24 hours. The circumference at the equator is about 24,000 miles, so any spot on the equator is moving at 1,000 miles per hour. Farther north, in New York City or San Francisco, the speed is slower: only about 760 miles per hour. A commercial jet airplane typically travels at 600 miles per hour.

If you're at a pole, you aren't moving but you are spinning—but only once per day—not enough to make you dizzy. Midlatitudes have both velocity and spin. It's the spinning of the Earth that makes hurricanes and cyclones spin.

For our even higher speed through the universe, see page 48.

If you are sitting still, you're actually traveling around the axis of the Earth faster than a jet airplane.

THE PHYSICS: What really killed the electric car was the price, which was super high because physicists and chemists failed to invent sufficiently cheap batteries. Rechargeable batteries last only 500 recharges. The cost of replacing them is much higher than the cost of the gasoline saved. Yes, some electric cars are being built, but replacing the batteries is so expensive that they are only for the wealthy (see page 16 for more on this).

Battery-driven cars can be practical if you can use cheap lead-acid batteries. But for the same energy, lead-acid batteries weigh 3–5 times more than lithium-ion batteries. Because of the limited space in a car, the range of a lead-acid car will be smaller, probably less than 50 miles between charges. That limited range may be acceptable in developing countries where the population is not yet addicted to long-range travel.

Maybe physicists and chemists will come to the rescue and develop a cheap, compact battery that can be recharged thousands of times. They're working on it.

Who killed the electric car? The oil industry?
Car companies? A conspiracy? No—
physics is the villain.

THE PHYSICS: The huge power transfer as you fill your car with gas is evident when you realize that in 3 minutes at the pump you can put enough energy in your car to drive a 2-ton vehicle 300 miles.

Let's look at the numbers. A gallon of gasoline can deliver about 120 million joules of heat energy, and it takes about 10 seconds to pump; that's 12 million joules per second. A watt is one joule per second, so you are pumping energy at a rate of 12 million watts, or 12 megawatts.

If we ever go to battery-powered cars (for example, plug-in hybrids), they will take much longer to recharge. The 15-amp circuit breaker used in most 110-volt homes allows the delivery of $15 \times 110 = 1,650$ watts. To match the refill time of a corner gasoline station (12 megawatts), that rate of energy delivery would have to be increased by about a factor of 7,300. But since electric energy can be used about 5 times more efficiently in a car than can gasoline heat energy, it will in fact take only 1,460 times longer. So instead of 100 seconds (to fill a 10-gallon tank of gasoline), it will take 146,000 seconds (about 40 hours) to charge up the battery. At a filling station you might use 10 such circuits, and then it would be only 4 hours to fill 'er up!

When you fill your car's gas tank, you're transferring energy at the rate of 12 megawatts! That's equal to the typical power delivered to 12,000 small houses.

THE PHYSICS: TNT is used as an explosive not because of its great energy, but because it can *release* that energy in a millionth of a second. That incredibly rapid release creates enormous pressure, enough to crack rock or concrete. We say that although cookies have greater *energy*, TNT has greater *power*. (The comparison is analogous to a long-distance runner having a greater *range*, but a sprinter having higher *speed*.)

A gram of TNT provides just a little less than one food calorie of energy, but a gram of chocolate chip cookies—the kind we love to bake full of chocolate, butter, and sugar—provides 5 food calories.

How can a cookie destroy a car? No—you don't set it off with a fuse, as in the illustration. Instead, feed a pound of cookies to some teenagers, give them sledgehammers, and tell them to go at it. Although the job won't be as quick as with TNT, it'll be more complete.

Want to destroy a car? A pound of chocolate
chip cookies delivers 6 times more energy
than a pound of TNT.

THE PHYSICS: When the sun is overhead, sunlight delivers a power of about one kilowatt per square meter. The best solar cells (that you can afford) convert 20% of that into useful electricity. That's 200 watts—about a quarter of a horsepower. To get even half of a horsepower you would need 2 square meters of solar cells, and you'd get that only if the 2 square meters were pointed directly at the sun, and if it wasn't cloudy.

Modern cars typically cruise at 20 horsepower and use 100 horsepower or more when passing. So your half-horsepower solar car would be pretty wimpy.

NASA uses some superefficient solar cells for its spacecraft. They are about twice as efficient as the commercially available ones, but they cost about $100,000 per square meter. That's worth doing for a billion-dollar spacecraft, but way too expensive for a car. And they would still deliver only one horsepower at most.

Want to drive a solar-powered car? You'll have to
be satisfied with an engine that delivers
less than one horsepower.

THE PHYSICS: Here's the physics of holograms and mirrors.

In a hologram, laser light hits the surface and makes electrons move. And the moving electrons emit light that makes us think there's something where there isn't.

In a mirror, light from the reflected object hits the surface and makes electrons move. And the moving electrons emit light that makes us think there's something where there isn't.

These are nearly identical explanations. So why aren't people as awed by mirrors as they are by holograms? Only because mirrors are so common. A few hundred years ago, good-quality mirrors were expensive and rare, and much more valued then they are today. The biblical phrase "through a glass darkly" (1 Corinthians 13, also the name of an Ingmar Bergman movie) refers to the poor image seen in ancient mirrors. Today's mirrors are miracles.

Modern mirrors are so good that a good hall of mirrors can completely confound. Mirrors are used by magicians who know that people can't distinguish a good reflection from reality. They produce a much better image than a hologram does. So next time you look in a mirror, take some joy at this amazing high-tech device.

Holograms seem exotic and mysterious.
Why don't we consider mirrors to be
exotic and mysterious?

THE PHYSICS: A terawatt is 1,000 gigawatts, roughly the power produced by 1,000 large power plants. That sounds like a lot, but global energy demand is now 14 terawatts. A terawatt of waves could provide only 7% of our needs, even if we used it all. For comparison, the solar power hitting land has about 50,000 terawatts.

Wave power may have value for local power needs, for towns close to the ocean. The difficulty is that it is expensive to maintain mechanical devices in the corrosive liquid known as seawater; ask anybody who tries to keep a boat shipshape. The world's largest wave power plant was, until recently, the Aguçadoura Wave Park operating 3 miles off the coast of Portugal. It had three 750-kilowatt power generators, each 460 feet (140 meters) long, producing a total of 2.25 megawatts of power. But the generators have been brought back to port because of problems with the bearings. For comparison, the largest wind turbine in Germany (see page 94) produces 8 megawatts from a single turbine, and a typical nuclear power generator produces 1,000 megawatts.

Waves hit the shores with about a terawatt of power. Sounds impressive. But even if we used it all as a source of renewable energy, that power is small compared to current human needs.